Toula 270

The Story of a Misunderstood Angle

Written by Sara Treem

Illustrations by Drew Dupre

Toula 270° was not a very happy angle. She always looked at the other kids in her class and noticed how different she was. They had great names and great looks, and she didn't even have a name; she just had a number.

There was Robby Right. He was a perfect angle who always did well in school

and got 90's on all of his tests. Teachers always said, "*Right* On, Robby!", or

"You've got the *right* idea!", whenever he answered correctly in class. He

wore the *right* kinds of clothes and always said the *right* things. He was

even *right* handed.

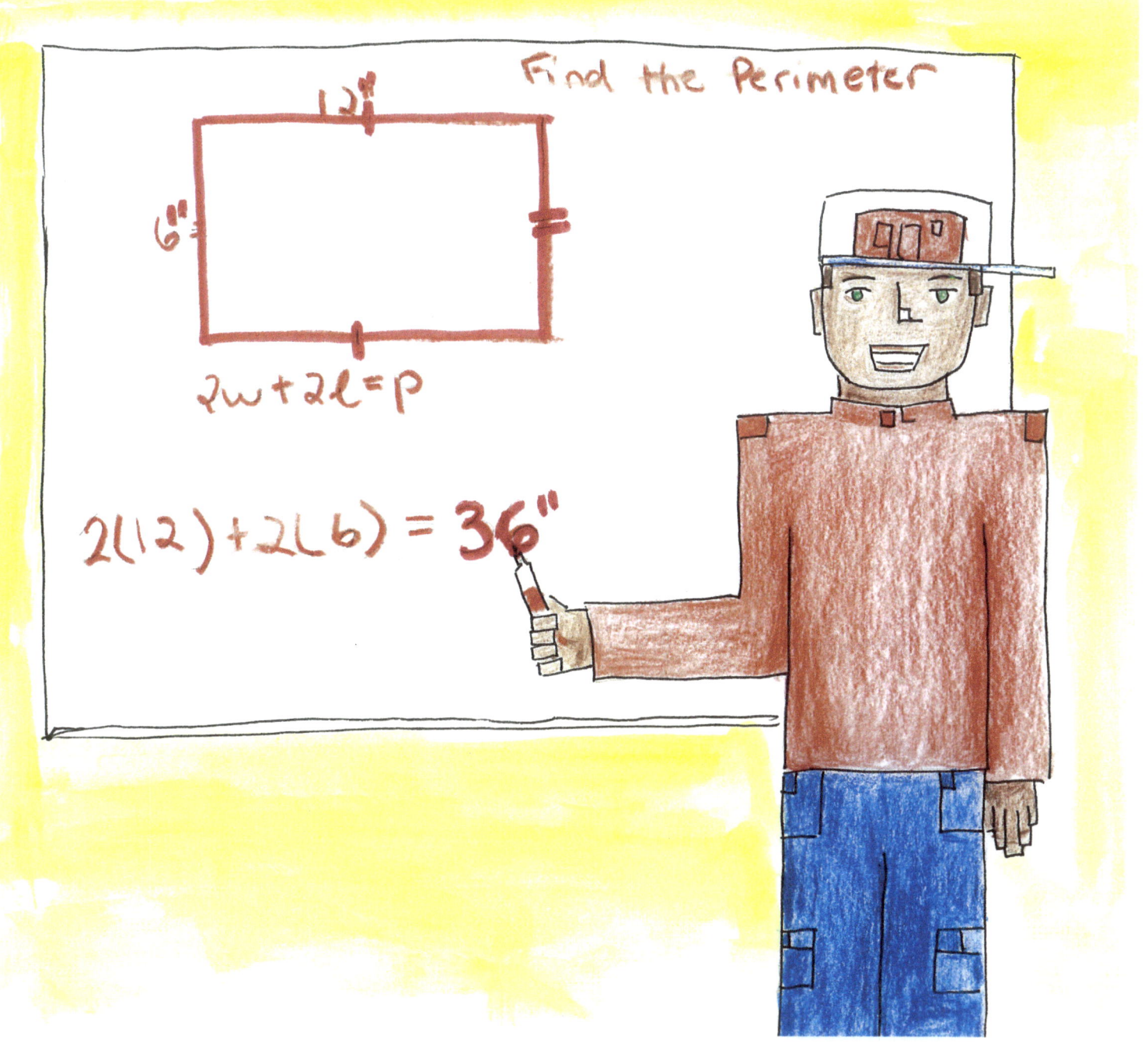

There was Sally Straight. She walked the straight and narrow. Everyone

liked her and teachers used her often as an example during math lessons.

She was also a great student and got *straight* As. Teachers liked her

because whenever they asked her for something, she'd respond *straight*

away.

There was C.C. Circle. He could be very silly at times, but was very good at completing tasks. He always seemed to be the one that made connections and could see things through all the way around. When a teacher was absent, he could be counted on to keep the ball rolling. He could talk around and around about any subject under the sun. If people were in a bad mood, he would say, "Oh, just roll with it."

There were the tiny, little Acute Buddies that everyone just adored.

Everything they did was soooo cute! They all looked a little different, but

were all considered acute.

There was the obtuse gang. They were so cool because they could spread

out and cover all the bases. Toula thought at first that maybe she belonged

with them. However, when Toula tried to approach the obtuse gang, she had

a problem. They said she couldn't join their gang and that she didn't fit in.

"You are not officially obtuse," Olly Obtuse said to her, "You are too big to join us. You are just so perfect! You are like three Robby Rights. Everyone needs to measure the rest of us with a protractor or an angle ruler first, but with you they just know. If you joined our gang you would just make us feel badly all the time."

Before she had a chance to defend herself he walked away and she didn't feel like joining his (or any) group anymore.

Then, one day, in Mr. Angle's math class something great happened that changed her life forever. Mr. Angle was their fifth grade teacher and he was talking about angles. He asked the class if they knew what "benchmark angles" were.

Everyone looked around to see if he was talking about a new kid or something. Mr. Angle was shocked and said how surprised he was that their other teachers hadn't taught them about how wonderful benchmark angles were!

Mr. Angle drew pictures of Robby Right, Sally Straight, and C.C. Circle on the board. He asked the kids to close their eyes and imagine these angles and how much they measured. The kids all knew that Robby was 90°, Sally was 180° and C.C. was 360°.

Next, Mr. Angle drew a picture of Toula 270° on the board.

All of the students looked around. No teachers had ever asked the class to

think of Toula as a special angle before.

"Close your eyes, class," said Mr. Angle. "Can you picture Toula?"

She started to get red in the face.

"Toula is a benchmark angle, just like Robby, Sally, and C.C. These are angles that can be easily distinguished. Toula is like three right angles. You know what they measure without having to use a protractor or an angle ruler. These angles are so useful! They help out mathematicians and scientists every day by giving them a reference point."

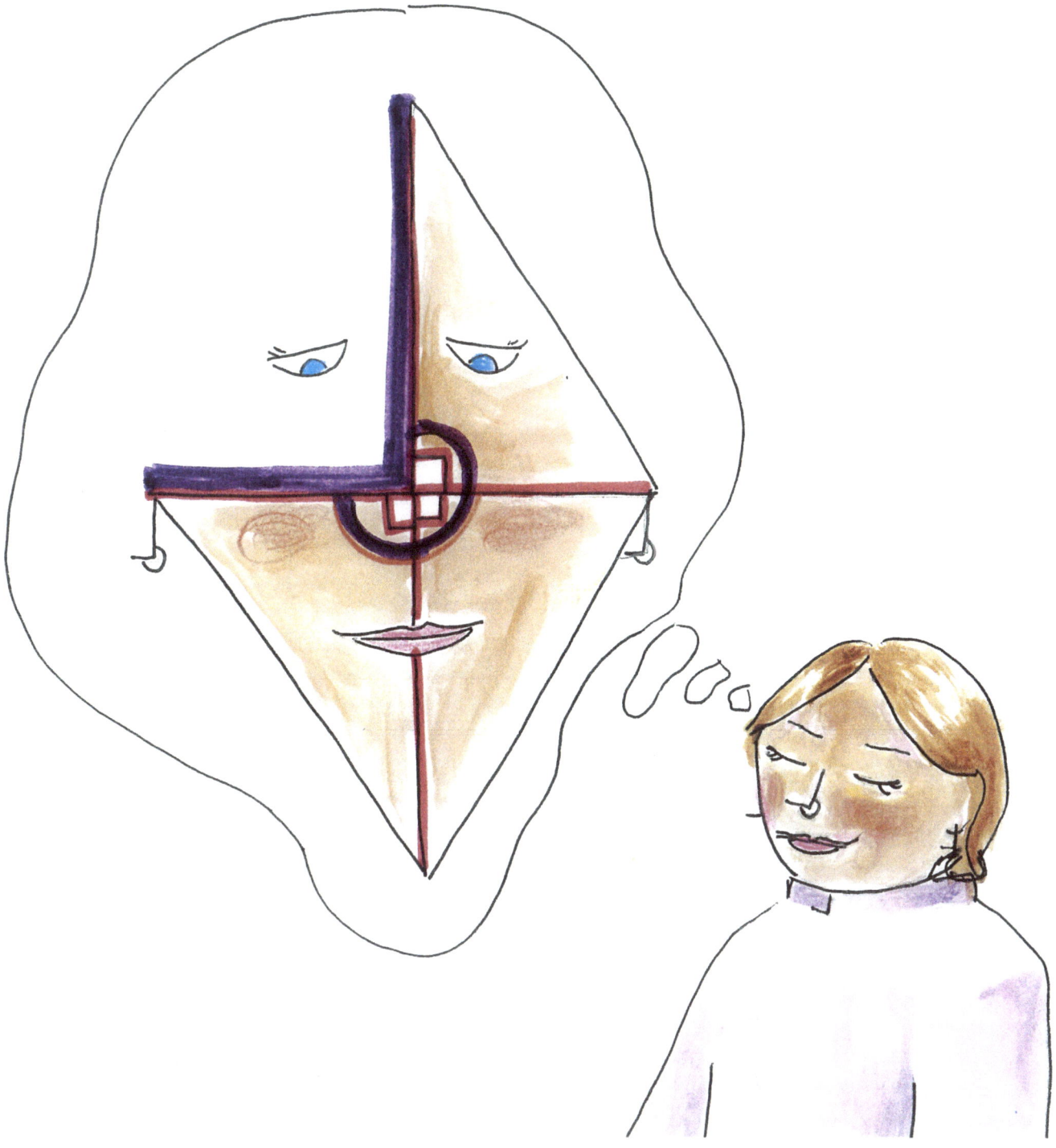

"Wow!" said all the kids.

"Toula is perhaps the most misunderstood benchmark angle of them all. She helps out with all of the angles that are no longer obtuse, but are not yet circles. These are called *reflex* angles. Thank goodness for Toula," said Mr. Angle.

At lunch, Robby, Sally, and C.C. wanted to form a table just for the benchmarks. Toula was supposed to sit at the head of the table. She decided that because benchmarks could be used by everybody that she would sit with everybody. She decided to take turns throughout the week and make friends with angles of all different shapes and sizes.

When Toula was at her locker, another reflex angle came up to talk to her.

"Hey Toula," he said, "Thanks for all of your help. Kids used to call me

'Reflex the Reject' and now they are saying how neat reflex angles are. I

know I'm not a benchmark like you, but it is still nice to know that there are

others like me."

"Any time!" said Toula.

For the rest of the day whenever people saw Toula they smiled at her and gave her a high five. They began to say, "Hey Benchmark, thank goodness for you," or "Benchmark 270°, you're terrific!"

She loved all the attention and the new nicknames. All of the nice things people said about her made her feel like she finally belonged.

For the rest of her life, she remained grateful to Mr. Angle. He showed her

that she was somebody. She was a benchmark angle. She was a cool 270°!

Toula 270 Math Glossary

acute angle: An angle which measures less than 90°.

angle ruler: An instrument used in measuring or drawing angles.

benchmark angle: A standard measurement that forms the basis for comparison.

circle: A 2-dimensional shape made by drawing a curve that is always the same distance from a center. It measures 360° all the way around.

obtuse angle: An angle which measures more than 90° but less than 180°.

protractor: An instrument used in measuring or drawing angles.

reflex angle: An angle which measures more than 180° but less than 360°.

right angle: An angle which measures exactly 90°.

straight angle: An angle which measures exactly 180°.

Other related terms not in this book:

complementary angles: Two angles which measure together to be 90°.

supplementary angles: Two angles which measure together to be 180°.

Note for Parents/Teachers

Here is a list of ideas for ways you could use this book with your children.

1. Before reading the book, discuss with children what they already know about angles. Do they know the names of special angles and what they measure? Do they know how to draw them? Etc.

2. Discuss real world connections, by asking children to explain ways angles are used in everyday life.

3. Distribute paper and protractors or angle rulers to children before reading. As you read the book, ask them to draw examples of the different angles. To check their work, have them swap the paper with another child.

4. Have children play a Tic Tac Toe-style game on a circular coordinate grid, like the one shown below. Players take turns calling out one coordinate at a time. The first player to get four in a row (along an arc *or* across a line) wins.

Note: Players may call the center of the grid by saying 0, 0°.

Example: *Player X:* Circle 1, 0°

   Player O: Circle 2, 90°

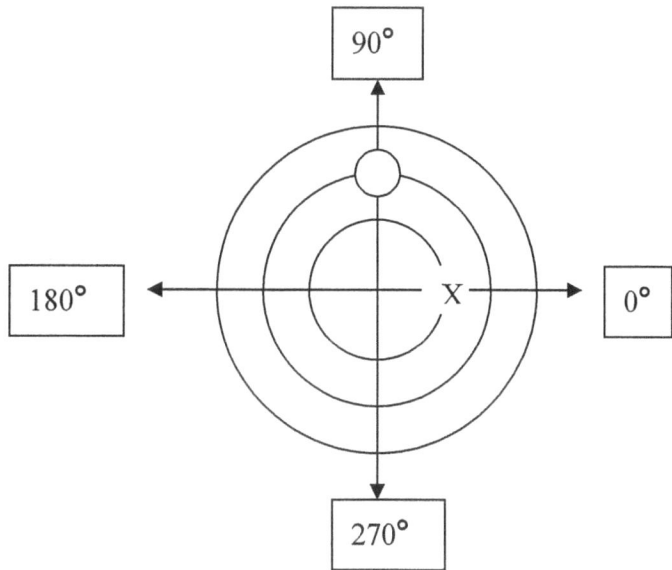